Please go to www.amathproblem.com for more types of problems.

In the book, you are going to see the term "the 3 elements formulas" : to learn how to use the 3 elements formulas to solve word problems, please go to amazon.com to order the books*: word problems: detailed explanations of reasoning and solving strategies* by Bill S. Lee; or go to www.amathproblem.com to find the introduction to each book first before you purchase it.

I0493842

The following 50 problems teach you how to use the least common multiples and the greatest common factors to solve word problems. If you need more similar problems to work on, please find volume 2. If you have any questions, please go to www.amathproblem.com to contact me

Rules to solve certain types of word problems on the least common multiples and the greatest common factors:

1. When the greatest common factor of two numbers divides into the two numbers, the quotients are prime to one another.

For example, the greatest common factor of 12 and 18 is 6. When 6 divides into 12 and 18, the quotients are 2 and 3: 2 and 3 are prime to each other.

2. The factors of the greatest common factors of numbers are also the factors of the numbers.

For examples, the greatest common factor of 18 and 12 is 6. 2 and 3 are the factors of 6; they are also the factors of 18 and 12.

3. The product of the greatest common factor and the least common multiple of two numbers equals to the product of the two numbers.

For example, the greatest common factor of 12 and 18 is 6; the least common multiple of 18 and 12 is 36. The product of 6 and 36 is 216; the product of 12 and 18 is also 216.

Using the above rules to solve the following problems

1. When a two-digit number divides into 50, the remainder is 2; when it divides into 63, the remainder is 3; when it divides into 73 the remainder is 1. What is the largest possible value of the 2-digit number?

2. How many factors does 24 have? How many factors does 36 have? What are the common factors of 24 and 36? What is the largest common factor of 24 and 36?

3. The product of two natural numbers is 420. If their

greatest common factor is 12, what is their least common multiple?

4. The least common multiple of two numbers is 126; the greatest common factor is 6. One of the numbers is 18, what is the other number?

5. The sum of two numbers is 48 and their greatest common factor is 6. What are the two numbers?

6. The GCF of two numbers is 15; the LCM of the numbers is 90. What are the two numbers?

7. The GCF of two numbers is 9; the LCM of them is 90. What

are the two numbers?

8. The GCF of two numbers is 60; the LCM of them is 720. If one of the numbers is 180, what is the other number?

9. The product of two numbers is 360 and the LCM of them is

120. What are the two numbers?

10. Find the product of the LCM and the GCF of 36 and 24.

11. The product of two numbers is 3,072. If their GCF is 16, what are the two numbers?

12. The GCF of two numbers is 13; their LCM is 78. Find the difference between the two numbers.

13. The GCF of two numbers is 4; the LCM of them is 252. If one of the numbers is 28, what is the other numbers?

14. When a number divides into 90, the remainder is 2; when the same number divides into 50, the remainder is also 2. What is the largest possible value of the number?

15. Find the numbers between 10 and 20 that are prime to 30; then find the sum of the numbers.

The Greatest Common Factors in Word Problems

16. A rectangular solid is 3.25 meters long, 1.75 meters wide and 0.75 meters high. It is cut into congruent cubes. What is the largest possible value of each edge of all the cubes and what is the largest number of cubes that the rectangular solid yields when the edges of the cubs have the largest possible value?

17. Mr. Lee gave all of 42 yellow balloons and all of 30 red balloons to groups of students and each group got the same number of yellow balloons and the same number of red balloons. At most how many groups got the balloons? How many of each did each group get?

18. There are 54 students in Mr. Smith's class and 63 students in Mr. Carl's class. The students in each class are divided into groups to go to the library. Each group in both classes has the same number of students in it. At most how many students are in each group? How many groups are in each class?

19. Students bought 24 lilies and 18 roses to make bouquets for their teachers. Each bouquet needs to have the same number of lilies and the same number of roses so at most how many bouquets can be made and how many lilies and roses are in each bouquet?

20. At most how many congruent squares can be cut out of a rectangle that is 84 inches long and 60 inches wide? How long will each side of the squares be?

21. One piece of wire is 18 inches long; the second piece of 24 inches long; the third piece is 30 inches long. If each of them is cut into sections and the length of each section is congruent, what is the largest possible value of each section? How many sections can we get?

22. A rectangle is 60 inches long and 36 inches wide. Congruent squares of the largest possible area are cut out of the rectangle. How long is each side of the squares? How many squares can be cut out of the rectangle?

23. Mom wants to make a certain number of bouquets with 96 red roses and 72 white roses. Each bouquet has the same number of red roses and the same number of white roses. At most how many bouquets can mom can? How many red rose and white roses are in each bouquet?

24. 24 watermelons and 36 cantaloupes are respectively divided into a certain number of groups. The number of watermelons in each watermelon group equals to the number of cantaloupes in each cantaloupes group. At most how many of each type of fruit are in each group? How many groups are the watermelons divided into? How many groups are the cantaloupes divided into?

25. There are 121 students in Team A and 143 students in Team B. Each team is divided into smaller groups and the number of students in each group is the same. At most how many students are in each group? How many groups is Team A divided into? How many groups is Team B divided into?

26. 320 pears, 240 peaches and 200 oranges are put in a certain number of baskets to make fruits baskets: each basket has the same number of peaches, the same number of pears and the same number of oranges. At most how many basketballs can all the fruits made? How many pears are in each basket? How many peaches are in each basket? How many oranges are in each basket?

27. A piece of rectangle paper that is 6 inches long and 4 inches wide is cut into the largest possible congruent squares. At most how many squares can we get? How long is each side of the squares?

28. A sheet of rectangle paper is 24 inches long and 15 inches wide. It is cut into congruent squares with no remaining paper. What is the largest possible side of each square? How many squares can be cut out of the rectangle sheet of paper?

29. Three ropes are 120 inches long, 90 inches long and 150 inches long. They are cut into sections with no remaining parts. If each section has the same length, at most how long is each section? How many sections can we get altogether?

The Least Common Multiples

30. Every 5 minutes, a bus station sends out Route 20; every 10 minutes the bus station sends out Route 30; every 6 minutes, the bus station sends out Route 40. At 5:00am the 3 buses are sent out together so in how much time will another group of the buses be sent out at the same time again?

31. Rectangular tiles that are 6 inches long and 4 inches wide are put on a floor and the tiles will make square patterns on the floor. What is the least possible length of each square pattern on the floor?

32. Mr. Smith has 3 sons: John, Peter and George. John visits Dad every 3 days; Peter visits Dad every 4 days; George visits Dad every 6 days. On July 1st, they all went home to visit Dad. At least how many days will pass before they go home on the same date again?

33. Both buses Route 1 and Route 2 are sent out of a bus station at 6:00am. After that time, Route 1 will be sent out every 45 minutes; Route 2 will be sent out every 1 hour. What is the lease amount of time needed before the two routes are sent out of the station at the same time again?

34. A certain number of tennis balls are packed into boxes: if each box has 12 in it, there will be 11 tennis ball left out; if each box has 18 in it, there will be 1 less for the last box; if each box has 15 in it, there will be 1 less for the last box. The number of tennis balls is between 300 and 400 so how many tennis balls are there to be packed?

35. There are less than 1,000 apples to be packed into boxes: if each box has 24 in it, there will be 2 less for the last box; if each box has 28 in it, there will be 2 less for the last box, too; if each box has 32 in it, the last box will have only 30 apples in it. How many apples are there altogether?

36. There are less than 1,000 tennis balls to be packed in boxes: whether they are packed by 4's, or 5's or 6's or 8's, there will always be 1 left out. How many tennis balls are there altogether?

37. When we put certain amount of oil in one type of bottles, there will be 3 liters short for the last bottle; when we put the oil in the second type of bottles the last bottle will be half full; when we put the oil in the third type of bottles, there will be 7 liters short for the last bottle. If the first type of bottle can hole 8 liters at most, the second type can hold 10 liters at most and the third type can hold 12 liters at most, at least how many liters of oil is there?

38. There are some marbles in a jar. When they are counted by 4's, there will be 3 left out; when they are counted by 6's, there will be 5 left out; when they are counted by 15's, there will be 14 left out. The number of marbles is between 150 and 200 so how many marbles are there in the jar?

39. When a certain number of tennis balls are counted by 9's there will be 7 left out; when they are counted by 10's there will be 8 left out; when they are counted by 12's, there will be 10 left out. If the number of tennis balls is between 150 and 200, how many tennis balls are there?

40. More than 50 but less than 60 students are divided into groups and each group has the same number of students. If they are divided into 4 groups, there will be 2 students left out; if they are divided into 5 groups, there will be 3 students left out. How many students are there?

41. Some apples are put in boxes. If each box has 30 put in it, there will be 20 apples left out; if each box has 35 put in it, there will be short of 10 apples for the last box. What is the least number of apples?

42. There are 25 light poles in a straight street and the distance between each is 45 yards. If we change the distance to 60 yards, how many poles do not need to be moved?

43. On the way from Peter's house to his friend's house there are 37 light poles. The distance between each is 50 yards. If the distance between each is 60 yards, how many poles do not need to be moved besides the first one and the last one?

44. There are 26 flags in a street and the distance between each is 4 yard. If the distance is changed to 5 yards, how many flags do not need to be moved besides the first one?

45. A certain number of tree seedlings are planted along a street and the distance between each seedling is 2 yards. The distance from the first tree seedling and the last one is 90 yards. When the seedlings grow bigger, some are moved so that the distance between each is 5 yards. How many tree seedlings do not need to be moved besides the first one and the last one?

46. Peter goes a library every 3 days; John goes to the same library every 4 days; George goes there every 5 days. Today, they all went to the library at the same. What is the least number of days that will pass before they meet again in the library?

47. Buses Route 1, Route 2 and Route 3 are sent out of a bus station at 7:00am every morning. After that time, Route 1 is sent out every 10 minutes; Route 2 is sent out every 15 minutes; Route 3 is sent out every 20 minutes. What is the least amount of time that needs to pass before the 3 routes are sent out at the same time again?

48. Peter, John and George started at the same time from the same starting point to run around track: Peter takes 120 seconds to run one lap around track; John needs 80 seconds to run one lap around track; George needs 100 seconds to run one lap around track. In how many seconds will the three people start at the same time from the starting point again?

49. Sarah visits Mom every 7 days; Hanna visits Mom every 6 days; Dana visits Mom every 14 days. They all went to visit mom on June 1st so in how many days will they visit Mom on the same day again?

50. A rectangular brick is 20 inches long, 12 inches wide and 6 inches thick. What is the largest possible number of bricks do we need to make a cube?

Answer Keys:

Please go to www.amathproblem.com for more types of problems.

In the book, you are going to see the term "the 3 elements formulas" : to learn how to use the 3 elements formulas to solve word problems, please go to amazon.com to order the books*: word problems: detailed explanations of reasoning and solving strategies* by Bill S. Lee; or go to www.amathproblem.com to find the introduction to each book first before you purchase it.

The following 58 problems teach you how to use the least common multiples and the greatest common factors to solve word problems.

Rules to solve certain types of word problems on the least common multiples and the greatest common factors:

1. When the greatest common factor of two numbers divides into the two numbers, the quotients are prime to one another.

For example, the greatest common factor of 12 and 18 is 6. When 6 divides into 12 and 18, the quotients are 2 and 3: 2 and 3 are prime to each other.

2. The factors of the greatest common factors of numbers are also the factors of the numbers.

For examples, the greatest common factor of 18 and 12 is 6. 2 and 3 are the factors of 6; they are also the factors of 18 and 12.

3. The product of the greatest common factor and the least common multiple of two numbers equals to the product of the two numbers.

For example, the greatest common factor of 12 and 18 is 6; the least common multiple of 18 and 12 is 36. The product of 6 and 36 is 216; the product of 12 and 18 is also 216.

Using the above rules to solve the following problems:

1. When a two-digit number divides into 50, the remainder is 2; when it divides into 63, the remainder is 3; when it divides into 73 the remainder is 1. What is the largest possible value of the 2-digit number?

Answer:

From "When a two-digit number divides into 50, the remainder is 2" we know that the number is a factor of 48; from "when it divides into 63, the remainder is 3" we know that the number is a factor of 60; from "when it divides into 73 the remainder is 1." We know that the number is a factor is 70. Now we need to find the common factor of 48, 60 and 70: 1, 2, 4, 6, 12. We need the largest possible value of the number so the two-digit number is 12.

2. How many factors does 24 have? How many factors does 36 have? What are the common factors of 24 and 36? What is the largest common factor of 24 and 36?

Answer:

24 has 8 factors: 1, 2, 3, 4, 6, 8, 12, 24

36 has 9 factors: 1, 2, 3, 4, 6, 9, 12, 18, 36

The common factors of 24 and 36 are: 1, 2, 3, 4, 6, 12

The greatest common factor is: 12

3. The product of two natural numbers is 420. If their greatest common factor is 12, what is their least common multiple?

Answer:

According to this rule: The product of the greatest common factor and the least common multiple of the two numbers equals to the product of the two numbers

The least common multiple of the two numbers is: $420 \div 12 = 35$

4. The least common multiple of two numbers is 126; the greatest common factor is 6. One of the numbers is 18, what is the other number?

Answer:

Please remember this rule:

The product of the greatest common factor and the least common multiple of the two numbers equals to the product of the two numbers

The product of the LCM and the GCF is: 126×6

The value of the other number is: $126 \times 6 \div 18 = 42$

5. The sum of two numbers is 48 and their greatest common factor is 6. What are the two numbers?

Answer:

Factor 48 and make sure that 6 is a factor:

$48 = 8 \times 6$

$48 = $ (one number + the other number) $\times 6$

One number + the other number = 8

Since 6 is the GCF, the two numbers in the parenthesis should be prime to each other so the numbers can be: 1, 7; 3, 5

Therefore, the two numbers should be:

$1 \times 6 = 6$ and $7 \times 6 = 42$; or $3 \times 6 = 18$ and $5 \times 6 = 30$

6. The GCF of two numbers is 15; the LCM of the numbers is 90. What are the two numbers?

Answer:

The product of the two numbers equals to the product of the GCF and the LCM so the product of the two numbers is:

15 ×90

15 ×90= 15 × the product of the other factors of one of the numbers ×15× the product of the other factors of the other number

Since their GCF is 15, the product of the other factors of one of the numbers is prime to the product of the other factors of the other number.

Therefore, the product of the other factors of one of the numbers can be 1, making one of the numbers 15, and the product of the other factors of the other number can be 6, making the other number 90; or the product of the other factors of one of the numbers can be 2, making one of the numbers 30, and the product of the other factors of the other number can be 3, making the other number 45.

7. The GCF of two numbers is 9; the LCM of them is 90. What are the two numbers?

Answer:

The product of two numbers equals to the product of their GCF and LCM so the product of the two numbers is:

9 ×90

9 ×90= 9 × the product of the other factors of one of the numbers ×9× the product of the other factors of the other number

Since their GCF is 9, the product of the other factors of one of the numbers is prime to the product of the other factors of the other number.

Therefore, the product of the other factors of one of the numbers can be 1, making one of the numbers 9, and the product of the other factors of the other number can be 10, making the other number 90; or the product of the other factors of one of the numbers can be 2, making one of the numbers 18, and the product of the other factors of the other number can be 5, making the other number 45.

8. The GCF of two numbers is 60; the LCM of them is 720. If one of the numbers is 180, what is the other number?

Answer:

The product of two numbers equals to the product of their GCF and LCM so the product of the two numbers is:

60×720

Since one number is 180 we divide this number into the

product of the two numbers to find the value of the other number:

$60 \times 720 \div 180 = 60 \times 4 = 240$

9. The product of two numbers is 360 and the LCM of them is 120. What are the two numbers?

Answer:

Since the product of two numbers equals to the product of their GCF and LCM the product of the GCF and LCM is also 360. Since the LCM is 120 the GCF is 3.

$360 = 3 \times$ the product of the other factors of one of the numbers $\times 3 \times$ the product of the other factors of the other number

Since 3 is their GCG, the product of the other factors of one of the numbers should be prime to the product of the other factors of the other number

Therefore, the product of the other factors of one of the numbers can be 1 and the product of the other factors of the other number can be 40. The first number is 3 and the second number is 120; or the product of the other factors of one of the numbers is 5, making one of the numbers 15, and the product of the factors of the other number is 8, making the second number 24.

10. Find the product of the LCM and the GCF of 36 and 24.

Answer:

Since the product of the two numbers equals to the product of their LCM and the GCF, the product of the LCM and the GCF should be:

24×36=864

11. The product of two numbers is 3,072. If their GCF is 16, what are the two numbers?

Answer:

Since the product of two numbers equals to the product of their LCM and their GCF the product of the CGF and the LCM is:

3072= 16 ×the product of the other factors of the first number ×16 ×the product of the other factors of the second number

Since 16 is their GCG the product of the other factors of one of the numbers is prime to the product of the other factors of the other number

Therefore, the product of the other factors of one of the numbers can be 1 and the product of the other factors of the other number can be 12. The two numbers can be 16 and 192. Also, the product of the other factors of one of the numbers can be 3 and the product of the other factors of the other number can be 4 so the two numbers can also be 48 and 64.

12. The GCF of two numbers is 13; their LCM is 78. Find the difference between the two numbers.

Answer:

The product of the two numbers equals to the product of the GCF and the LCM:

The product of the two numbers=13 ×78= 13 ×the product of the other factors of one of the numbers ×13 × the product of the other factors of the other number

Since 13 is their GCF, the product of the other factors of one of the numbers is prime to the product of the other factors of the other number. The product of the other factors of one of the numbers can be 1, making one of the numbers 13, and the product of the other factors of the other number can be 78, making the other number 1,014. Or, the product of the other factors of one of the numbers can be 2, making the number 26, and the product of the other number can be 39, making the other number 507.

The difference between the two numbers can be:

1014−13=1,001

Or:

507−26=481

13. The GCF of two numbers is 4; the LCM of them is 252. If one of the numbers is 28, what is the other numbers?

Answer:

Since the product of the two numbers equals to the product of their LCM and GCF the product of the two numbers is:

4×252

Since one of the numbers is 28, the other number should be:

4×252÷28=1008÷28=36

14. When a number divides into 90, the remainder is 2; when the same number divides into 50, the remainder is also 2. What is the largest possible value of the number?

Answer:

When the number divides into 90, the remainder is 2, the

number should be a divisor of 88 (90-2).

When the number divided into 50 and the remainder is 2, the number should be a divisor of 48 (50-2).

Therefore, we are looking for a common divisor or a common factor.

Find the greatest common factor of 88 and 48: 8

Therefore, the number is 8.

15. Find the numbers between 10 and 20 that are prime to 30; then find the sum of the numbers.

Answer:

The numbers that are prime to 30 are: 11, 13, 17, and 19.

The sum of the 4 numbers is: 60

The Greatest Common Factors in Word Problems (GCF)

16. A rectangular solid is 3.25 meters long, 1.75 meters wide and 0.75 meters high. It is cut into congruent cubes. What is the largest possible value of each edge of all the cubes and what is the largest number of cubes that the rectangular solid yields when the edges of the cubs have the largest possible value?

Answer:

Since each edge of the cubes is the same, we need to find the common factors for the length, the width and the height of the rectangular solid. Since we want the edges to have the largest possible value, we need to find the larges common factors of the length, the width and the height of the rectangular solid.

What is the largest common factor for 3.25, 1.75 and 0.75? We need to first convert the decimals into whole numbers: 3.25 meters=325 centimeters

1.75 meters=175 centimeters

0.75 meters=75 centimeters

The largest common factor of 325, 175 and 75 is 25. Therefore the edge of each cube is 25 centimeters long or is 0.25 meters long.

Let's now use the 3 elements formulas to find how many cubes can be cut out of the rectangular solid:

The volume of each cube: 25 x 25 x 25

The volume of the rectangular solid: 325 x 175 x 75

The volume of the rectangular solid ÷ the volume of each cube= 273

So the rectangular solid can yield 273 cubes.

17. Mr. Lee gave all of 42 yellow balloons and all of 30 red balloons to groups of students and each group got the same number of yellow balloons and the same number of red balloons. At most how many groups got the balloons? How many of each did each group get?

Answer:
Do we need to find the greatest common factor or the least common multiple?
In this problem, one of the 3 elements is known: the total amount in all the groups: the total amount of yellow balloons is 42; the total number of red balloons is 30, so we need to find the number of groups and the amount in each group. To find the number of groups and the amount in each group is to find the factors of a numbers3

Since each group has the same number of yellow balloons, we need to see how many groups 42 balloons can be divided into: 42 can be divided into many groups: we need to find the factors of 42: 2, 21, 3,7, 6, 21

With the same reasoning: 30 can be divided into many groups: we need to find the factor of 30: 2, 3, 5, 6, 15, and 30

The number of groups of students that got the yellow balloons equals to the number of groups that got the red balloons so we need to the greatest common factor of 42 and 30: it is 6 so at most 6 groups of students got the balloons.

Using the 3 elements formulas, the number of yellow balloons each group got was:

$42 \div 6 = 7$

With the same reasoning, the number of red balloons each group got was:

$30 \div 6 = 5$

Each group got 7 yellow balloons and 5 red balloons.

18. There are 54 students in Mr. Smith's class and 63 students in Mr. Carl's class. The students in each class are divided into groups to go to the library. Each group in both classes has the same number of students in it. At most how many students are in each group? How many groups are in each class?

Answer:

Since each group has the same number of students, we need to find the amount in each group in the 3 elements, which means that we need to find the factors of 54 and 63:

In Mr. Smith's class, the possible number of students in each group is the factors of 54:1, 2, 3, 6, 9, 18, 27 and 54

In Mr. Carl's class, the possible number of students in each group is the factors of 63:1, 3, 7, 9, 21, 63

Since each group in both classes has the same number of students and since we need to find the larges possible number of students in each group, we need to find the greatest common factor of 54 and 63, which is 9.

According the 3 elements formulas:
The number of groups in Mr. Smith's class should be: $54 \div 9 = 6$
The number of groups in Mr. Carl's class should be: $63 \div 9 = 7$

19. Students bought 24 lilies and 18 roses to make bouquets for their teachers. Each bouquet needs to have the same number of lilies and the same number of roses so at most how many bouquets can be made and how many lilies and roses are in each bouquet?

Answer:
In other words, the number of bouquets the have lilies should equal to the number of bouquets that have roses so we need to find the number of groups in the 3 elements formulas. to find the number of groups is to find the factors.
The factors of 24: 1, 2, 3, 4, 6, 8, 12, 24
The factors of 18: 1, 2, 3, 6, 9, 18
Since we have the same number of bouquets or the same groups of lilies and roses, and we want the largest possible number of groups, we need the largest common factor of 24 and 18, which is 6.
So there are 6 bouquets and each bouquets have the same number of lilies and the same number of roses.
According to the 3 elements formulas:
The number of lilies in each bouquet is: $24 \div 6 = 4$
The number of roses in each bouquet is: $18 \div 6 = 3$

20. At most how many congruent squares can be cut out of a rectangle that is 84 inches long and 60 inches wide? How long will each side of the squares be?

Answer:
In other words, the amount in each group or each side of a square from 84 and 60 should be the congruent; we need to find the common factors of 84 and 60. Since we want the sides to have the largest possible value, we need to find the greatest common factor of 84 and 60, which is 12 inches.
Each side of a square is 12 inches so the area of it should be: 12×12
The area of the rectangle should be: 84×60
How many squares are in the rectangle? Use the 3 elements to find it:
$84 \times 60 \div (12 \times 12) = 35$

21. One piece of wire is 18 inches long; the second piece of 24 inches long; the third piece is 30 inches long. If each of them is cut into sections and the length of each section is congruent, what is the largest possible value of each section? How many sections can we get?

Answer:

Since the length of each section is the same, which means the amount in each group is the same, and since we need to find the length of each section, we need to find the factors of 18, 24 and 30. Since we want the largest possible value of each

44

section, we are looking for the greatest common factor of 18, 24 and 30: it is 6.

Each section is 6 inches long.

Use the 3 elements formulas to find the number of sections each piece of wire is cut into:

18÷6=3 sections

14÷6=4 sections

30÷6= 5 sections

3+4+5=12 sections

We can get altogether 12 sections and each section is 6 inches long.

22. A rectangle is 60 inches long and 36 inches wide. Congruent squares of the largest possible area are cut out of the rectangle. How long is each side of the squares? How many squares can be cut out of the rectangle?

Answer:

The largest possible value of the area means that each side of the square has its largest possible value. We need to find the greatest common factor of 60 and 36, which is 12 inches.

The area of each square is: 12×12

The area of the rectangle is: 60 × 36

Using the 3 elements formulas to find the number of squares that can be cut out of the rectangle:

60 ×36 ÷ (12 ×12)=15

23. Mom wants to make a certain number of bouquets with 96 red roses and 72 white roses. Each bouquet has the same number of red roses and the same number of white roses. At most how many bouquets can mom can? How many red rose and white roses are in each bouquet?

Answer:

How many bouquets can the red roses make? We need to find the factors of 72; how many bouquets can 96 red roses make? We need to find the factors of 96. Since the number of bouquets that the red roses are in equals to the number of bouquets that the number of white roses is in, we need to find the common factors of 72 and 96. Since we need the largest possible number of bouquets, we need the greatest common factor of 72 and 96, which is 24.

Using the 3 elements formulas, find the number of red roses in each bouquet:

96÷24=4 red roses

Using the 3 elements formulas, find the number of white roses in each bouquet:

72÷24=3 white roses

24. 24 watermelons and 36 cantaloupes are respectively divided into a certain number of groups. The number of watermelons in each watermelon group equals to the number of cantaloupes in each cantaloupes group. At most how many of each type of fruit are in each group? How many groups are the watermelons divided into? How many groups are the cantaloupes divided into?

Answer:

Since the number of watermelons in each watermelon group equals to the number of cantaloupes in each cantaloupe group, we need to find the common factors of 24 and 36; since we need the largest possible number of watermelons in each group and the largest possible number of cantaloupes in each group, we need to find the greatest common factors of 24 and 36, which is 12.

Using the 3 elements formulas, the number of groups the watermelons are divided into is:

24÷12=2

Using the 3 elements formulas the number of groups that the cantaloupes are divided into is:

$36 \div 12 = 3$ groups

25. There are 121 students in Team A and 143 students in Team B. Each team is divided into smaller groups and the number of students in each group is the same. At most how many students are in each group? How many groups is Team A divided into? How many groups is Team B divided into?

Answer:

Since "the number of students in each group is the same." We need to find the common factor of 121 and 143. Since we need to find "at most" how many students are in each group, we need to find the greatest common factor of 121 and 143, which is 11.

Using the 3 elements formulas, we can find the number of groups Team A is divided into:

$121 \div 11 = 12$

Using the 3 elements formulas, we can find the number of groups Team B is divided into:

$143 \div 11 = 13$

26. 320 pears, 240 peaches and 200 oranges are put in a

certain number of baskets to make fruits baskets: each basket has the same number of peaches, the same number of pears and the same number of oranges. At most how many basketballs can all the fruits made? How many pears are in each basket? How many peaches are in each basket? How many oranges are in each basket?

Answer:

The number of baskets that the pears are in not only equals to the number of baskets that peaches are in but also equals to the number of baskets the oranges are in. In other words, the pears, peaches and oranges are each divided into groups and the number of groups of each type of fruit is divided into is equal. When dividing a number into groups to find the number of groups, we look for the factors of the number. Since each fruit is in the same number of groups as the others, in other words, in the same number of baskets, we need to find the common factors of 320, 240 and 200. Since we want the largest possible number of baskets, we need to find the greatest common factor of 320, 240 and 200, which is 40. At most 40 fruit baskets can be made from the fruits.

Using the 3 element formulas, we find the number of each type of fruits in each basket:

$320 \div 40 = 8$ pears

$240 \div 40 = 6$ peaches

$200 \div 40 = 5$ oranges

27. A piece of rectangle paper that is 6 inches long and 4 inches wide is cut into the largest possible congruent squares. At most how many squares can we get? How long is each side of the squares?

Answer:

A square has 4 equal sides so we need to find common factors of the length and the width. Since we want the largest possible squares, we need to find the greatest common factors of 6 and 4, which is 2 inches.

The area of each square is: $2 \times 2 = 4$ square inches

The area of the rectangle is: $6 \times 4 = 24$ square inches

Using the 3 elements formulas to find the number of squares that can be cut out of the rectangle paper:

$24 \div 4 = 6$ squares

28. A sheet of rectangle paper is 24 inches long and 15 inches wide. It is cut into congruent squares with no remaining paper. What is the largest possible side of each square? How many squares can be cut out of the rectangle sheet of paper?

Answer:

Since each side of a square is equal, we need to find the greatest common factor of 24 and 15, which is 3. Therefore, each square is 3 inches long. The area of each square is 9 square inches.

The area of the rectangle is 24 ×15=360 square inches

Using the 3 elements formulas to find the number of squares in the rectangle:

360÷9=40

40 squares can be cut out of the rectangle.

29. Three ropes are 120 inches long, 90 inches long and 150 inches long. They are cut into sections with no remaining parts. If each section has the same length, at most how long is each section? How many sections can we get altogether?

Answer:

To cut the ropes into sections is like to divide each rope into different groups and each group has the same amount in it. We are supposed to find the amount in each section. Finding the amount in each group is to find a factor of a number which in this question is the length of each rope. Since each section has the same length, we want to find the greatest (at most how long) common factor of the three numbers 120, 90 and 150, which is 30 inches.

The rope that is 120 inches long is cut into 4 sections.

The rope that is 90 inches long is cut into 3 sections.

The rope that is 150 inches long is cut into 5 sections.

Therefore, all the ropes are cut into 12 sections.

The Least Common Multiples (LCM)

30. Every 5 minutes, a bus station sends out Route 20; every 10 minutes the bus station sends out Route 30; every 6 minutes, the bus station sends out Route 40. At 5:00am the 3 buses are sent out together so in how much time will another group of the buses be sent out at the same time again?

Answer:

In another 5 minutes, which is 5:05, another Route 20 will be sent out; in another 10 minutes, which is 5:10, the third Route 20 will be sent out. Bus Route 20 will be sent out in this patter every 5 minutes. From the pattern, we know that the 5 minutes, 10 minutes, ··· these are all the multiples of 5.

In another 10 minutes, which is 5:10, another Route 30 will be sent out; in another 20 minutes, which is 5:20, the third but Route 30 will be sent out. In this pattern, we know that 10, 20··· they are all multiples of 10.

With the same reasoning: for Bus Route 40, we also need to find the multiples of 6 to find different times at which Bus Route 40 is sent out.

When is the next time when all three routes are sent out again at the same time since 5:00? We need to find the least common multiple of 10, 5 and 6, which is 30. Therefore, in 30 minutes, which is at 5:00, the three routes of buses will be sent out at the same time again.

31. Rectangular tiles that are 6 inches long and 4 inches wide are put on a floor and the tiles will make square patterns on the floor. What is the least possible length of each square pattern on the floor?

Answer:

When we put two tiles on the floor, each side of the patterns on the floor should be multiples of 6 and 4. Since we want square patterns on the floor, we need to find the common multiple of 6 and 4. Since we need the least possible of length of each square, we need to find the least common multiple of 6 and 4, which is 12.

Basically, every 6 rectangle tiles make one square pattern.

32. Mr. Smith has 3 sons: John, Peter and George. John visits Dad every 3 days; Peter visits Dad every 4 days; George visits Dad every 6 days. On July 1st, they all went home to visit Dad. At least how many days will pass before they go home on the same date again?

Answer:

For John, the next few times for him to visit Dad will be: 3 days later, 6 later, 9 days later··· which are all multiples of 3.

For Peter, the next few times for him to visit Dad will be: 4 days later, 8 days later, 12 days later··· which are all multiples of 4.

For George, the next few times for him to visit Dad will be: 6 days later, 12 days later, 18 days later··· which are all multiples of 6.

The same number of days needs to pass before they visit Dad on the same day again so we need to find the least common multiple for 3, 4 and 6, which is 12. In 12 days, which is July 13th, they all will visit Dad again.

33. Both buses Route 1 and Route 2 are sent out of a bus station at 6:00am. After that time, Route 1 will be sent out every 45 minutes; Route 2 will be sent out every 1 hour. What is the lease amount of time needed before the two routes are sent out of the station at the same time again?

Answer:

For Route 1: the next few buses to be sent out will be 45 minutes later, 90 minutes later, 135 minutes later··· which are all the multiples of 45.

For Route 2: the next few buses to be sent out will be 60 minutes later, 120 minutes later, 180 minutes later··· which are all the multiples of 60.

The same amount of time has to pass before the two routes of buses are sent out at the same time again, so we need to find the least (the lease amount of time) common multiple of 45 and 60, which is 180. At 9:00am, Route 1 and Route 2 will be sent out.

34. A certain number of tennis balls are packed into boxes: if each box has 12 in it, there will be 11 tennis ball left out; if each box has 18 in it, there will be 1 less for the last box; if each box has 15 in it, there will be 1 less for the last box. The number of tennis balls is between 300 and 400 so how many tennis balls are there to be packed?

Answer:

From "if each box has 12 in it, there will be 11 tennis ball left out" in other words, if each box has 12 in it, the last box should have 1less so the number of tennis balls should be a multiple of 12 minus 1.

From "if each box has 18 in it, there will be 1 less for the last box;" we know that the number of tennis balls should be a multiple of 18 minus 1.

With the same reason, the number of tennis balls should also be a multiple of 15 minus 1.

We need to find the common multiple of 12, 18 and 15 and deduct 1 from the multiple: let's first find the least common multiple of 12, 18 and 15:180. Since "The number of tennis balls is between 300 and 400" the multiple of the three numbers has to be: 180 ×2=360

The number of tennis balls should be: 360-1=359

35. There are less than 1,000 apples to be packed into boxes: if each box has 24 in it, there will be 2 less for the last box; if each box has 28 in it, there will be 2 less for the last box, too; if each box has 32 in it, the last box will have only 30 apples in it. How many apples are there altogether?

Answer:

If we add 2 apples to the available apples, the number of apple should be a multiple of 24, 28 and 32. Let's first find the multiple of the three numbers and deduct 2 from it.

It's always easy to find the least common multiple first and if needed, we can find the larger multiples.

The least common multiple of 24, 28 and 32 is 672. Since there are less than 1000 apples, 672 should be the multiple that we will use: the next multiple will be 1,344.

672-2=670 apples

36. There are less than 1,000 tennis balls to be packed in boxes: whether they are packed by 4's, or 5's or 6's or 8's, there will always be 1 left out. How many tennis balls are there altogether?

Answer:

The number of tennis balls should be a common multiple of 4, 5, 6, 7 and 8, plus 1. Let's first find the least common multiple of the numbers: the final answer does not have to be the least common multiple, but it is easy to find a common multiple of numbers when we find the least common multiple first.

The least common multiple of 4, 5, 6, 7, and 8 is :840. Since the number of tennis balls is less than 1,000 we will use this number:

The number of tennis balls is: 840+1=841

37. When we put certain amount of oil in one type of bottles, there will be 3 liters short for the last bottle; when we put the oil in the second type of bottles the last bottle will be half full; when we put the oil in the third type of bottles, there will be 7 liters short for the last bottle. If the first

type of bottle can hole 8 liters at most, the second type can hold 10 liters at most and the third type can hold 12 liters at most, at least how many liters of oil is there?

Answer:

we can reword the question in this way: 8 liters are put in each bottle, there will be 5 liters too much (8-3=5); when 10 liters are put in each bottle, there will be 5 liters too much (half of 10 is 5); when 12 litters are put in each bottle, there will be 5 liters too much (12-7=5). Therefore, the amount of oil has to be a common multiple of 8, 10 and 12 plus 5. Since we need the least possible amount of oil, we need to find the least common multiple of 8, 10 and 12, which is 120. The amount of oil is:

120+5=125 liters

38. There are some marbles in a jar. When they are counted by 4's, there will be 3 left out; when they are counted by 6's, there will be 5 left out; when they are counted by 15's, there will be 14 left out. The number of marbles is between 150 and 200 so how many marbles are there in the jar?

Answer:

Let's reword the question: when counted by 4's, there will

be 1 short; when counted by 6' s, there will be 1 short; one counted by 15' s, there will be 1 short. Therefore, the number of marbles is a common multiple of 4, 6 and 15 minus 1.

Let's find the least common multiple of the 3 numbers first: 60. Since the number of marbles is between 150 and 200, the number of marbles has to be a multiple of 60: 60 ×3=180

The number of marbles is: 180−1=179

39. When a certain number of tennis balls are counted by 9' s there will be 7 left out; when they are counted by 10' s there will be 8 left out; when they are counted by 12' s, there will be 10 left out. If the number of tennis balls is between 150 and 200, how many tennis balls are there?

Answer:

Let's reword the question: when they are counted by 9' s, there will be 2 short; when they are counted by 10' s , there will be 2 short; when they are counted by 12' s, there will be 2 short. Therefore, the number of tennis balls is a multiple of 9, 10 and 10 minus 2.

The least common multiple of 9, 10 and 12 is 180. Since the number of tennis balls is between 150 and 200, we will use the least common multiple: deduct 2 from 180: 180−2=178

There are 178 tennis balls there.

40. More than 50 but less than 60 students are divided into groups and each group has the same number of students. If they are divided into 4 groups, there will be 2 students left out; if they are divided into 5 groups, there will be 3 students left out. How many students are there?

Answer;

Let's reword the question: when the students are divided into 4 groups, there will be short of 2 students; when they are divided into 5 groups, there will be short of 2 students. Therefore, we need to find a common multiple of 4 and 5 minus 2. The least common multiple of 4 and 5 is 20. Since the number of students is between 50 and 60, we need to use a multiple of 20 to solve the problem:

20 ×3=60

60 is still a common multiple of 4 and 5.

The number of students is: 60−2=58

41. Some apples are put in boxes. If each box has 30 in it, there will be 20 apples left out; if each box has 35 in it, there will be short of 10 apples for the last box. What is the least number of apples?

Answer:

Let's reword the second sentence: if each box has 30 in it, there will be short of 10 for the last box. Combining the reworded second sentence and the third sentence, we know that the number of apples is the least (the least possible number of) common multiple of 30 and 35 minus 10. The least common multiple of 30 and 35 is 210

The number of apples is: 210-10=200

42. There are 25 light poles in a straight street and the distance between each is 45 yards. If we change the distance to 60 yards, how many poles do not need to be moved?

Answer:

The light poles that do not need to be moved are those whose distance from the first light pole is a common multiple of 45 and 60. Let's first find the least common multiple of 45 and 60, which is 180.

How long is the entire street that has the 25 light poles in?

Using the 3 elements formals to find the distance from the first light pole to the last one:

45 × (25-1) =1,080 yards

Every 180 yards, the light poles do not need to be moved.

Using the 3 elements formulas to find how many light poles do not need to be moved:

$1080 \div 180 = 6$

The first one does not need to be moved so the total number of light poles that do not need to me moved is: 6+1= 7

43. On the way from Peter's house to his friend's house there are 37 light poles. The distance between each is 50 yards. If the distance between each is 60 yards, how many poles do not need to be moved besides the first one and the last one?

Answer:

The light poles that do not need to be moved are those whose distance from the first light pole is a common multiple of 50 and 60. Let's first find the least common multiple of 50 and 60, which is 300.

How long is the entire street that has the 37 light poles in?

Using the 3 elements formals to find the distance from the first light pole to the last one:

$50 \times (37-1) = 1,800$ yards

Every 300 yards, the light poles do not need to be moved. Using the 3 elements formulas to find how many light poles do not need to be moved:

$1800 \div 300 = 6$

The 6 light poles include the last one so we deduct 1 from 6 to find how many do not need to be moved:

$6-1=5$

44. There are 26 flags in a street and the distance between each is 4 yard. If the distance is changed to 5 yards, how many flags do not need to be moved besides the first one?

Answer:

The flags that do not need to be moved are those whose distance from the first flag is a common multiple of 4 and 5. Let's first find the least common multiple of 4 and 5, which is 20. .

How long is the entire street that has the 26 flags in?

Using the 3 elements formals to find the distance from the first flag to the last one:

$4 \times (26-1) = 100$ yards

Every 20 yards, the flags do not need to be moved. Using the 3 elements formulas to find how many flags do not need to be moved:

$100 \div 20 = 5$

Besides the first one, another 5 flags do not need to be moved.

45. A certain number of tree seedlings are planted along a street and the distance between each seedling is 2 yards. The distance from the first tree seedling and the last one is 90 yards. When the seedlings grow bigger, some are moved so that the distance between each is 5 yards. How many tree seedlings do not need to be moved besides the first one and the last one?

Answer:

The tree seedlings that do not need to be moved are those whose distance from the first tree seedling is a common multiple of 2 and 5. Let's first find the least common multiple of 2 and 5, which is 10.

Every 10 yards, the tree seedlings do not need to be moved. Using the 3 elements formulas to find how many tree seedlings do not need to be moved:

$90 \div 10 = 9$

The last one is included in the 9 seedlings so we need to deduct 1 from 9 to find how many tree seedlings do not need to be moved besides the first one and the last one:

46. Peter goes a library every 3 days; John goes to the same library every 4 days; George goes there every 5 days. Today, they all went to the library at the same. What is the least number of days that will pass before they meet again in the library?

Answer:

The next few times when Peter goes to the library will be 3 days later, 6 days later, 9 days later… which are all multiples of 3.

The next few times when John goes to the library will be 4 days later, 8 days later, 12 days later…which are the multiples of 4.

With the same reasoning, the next few times when George goes to the library will be the multiples of 5.

The same number of days needs to pass before the 3 meet again. Since we need to first the least possible number of days to pass before they meet again, we need to find the least common multiple of 3, 4, and 5, which 60.

In another 60 days they will meet again in the library.

47. Buses Route 1, Route 2 and Route 3 are sent out of a bus station at 7:00am every morning. After that time, Route 1 is sent out every 10 minutes; Route 2 is sent out every 15 minutes; Route 3 is sent out every 20 minutes. What is the least amount of time that needs to pass before the 3 routes are sent out at the same time again?

Answer:

The next few times when Route 1 is sent out will be in 10 minutes, 20 minutes, 30 minutes⋯ these are the multiples of 10.

The next few times when Route 2 is sent out will be in 15 minutes, 30 minutes, 45 minutes⋯these are the multiples of 15.

With the same reasoning, the next fewer times when Route 3 is sent out will be in 20 minutes, 40 minutes⋯ which are the multiples of 20.

We need to find the least common multiple of 10, 15 and 20 to find the least amount time to pass before the 3 routes are sent together.

The least common multiple of 10, 15 and 20 is 60. In another 60 minutes the 3 routes will be sent out of the same bus stations at the same time.

48. Peter, John and George started at the same time from the same starting point to run around track: Peter takes 120 seconds to run one lap around track; John needs 80 seconds to run one lap around track; George needs 100 seconds to run one lap around track. In how many seconds will the three people start at the same time from the starting point again?

Answer:

Peter takes 240 seconds to run the second lap, 360 seconds to run the third trap, 480 seconds to run the fourth lap… these are all the multiples of 120

John takes 160 seconds to run the second lap; 240 seconds to run the third lap… there are the multiples of 80

George takes 200 seconds to run the second lap; 300 seconds to run the third lap…there are the multiples of 100.

We need to find the least common multiple of 120, 80 and 100 to find the amount of time the three people need to start at the same time from the same starting point again. The least common multiple of 120, 80 and 100 is 1,200.

In 1,200 seconds or in 20 minutes they will start from the same starting point again.

49. Sarah visits Mom every 7 days; Hanna visits Mom every 6

days; Dana visits Mom every 14 days. They all went to visit mom on June 1ˢᵗ so in how many days will they visit Mom on the same day again?

Answer:

The next times when Sarah visits Mom will be in 7 days, 14 day, 21 days···these are all multiples of 7

The next times when Hanna visits Mom will be in 6 days, 12 days, 18 days···these are all multiples of 6

The next times when Dana visits Mom will be in 14 days, 28 days, 42 days···these are the multiple of 14.

We need to find the least common multiple of 7, 6 and 14 to find the next same date when they visit Mom again.

The least common multiple of 7, 6 and 14 is 42 so in 42 days they will visit Mom on the same day again.

50. A rectangular brick is 20 inches long, 12 inches wide and 6 inches thick. What is the largest possible number of bricks do we need to make a cube?

Answer:

Each edge of a cube is equal so we need to find the least common multiple of 20, 12 and 6 to find the edge of each cube. The least common multiple of 20, 12 and 6 is 60. Each edge of a cube is 60 inches so the volume of each cube is: 216,000 cubic inches

The volume of each rectangle is: $20 \times 12 \times 6 = 1,440$ cubic inches

Using the 3 elements formulas to find the number of rectangular bricks needed to make a cubic brick:

$216,000 \div 1,440 = 150$

www.ingramcontent.com/pod-product-compliance
Lightning Source LLC
Chambersburg PA
CBHW081851170526
45167CB00007B/2971